Lah'aj Kids Multiplication Worksheets
With Ṡabaṫiyya

By Tiyi Hibner
©2025 Yamartat Ta'/Tiyi Hibner
Published by www.TempleBabies.com

ṢABAṪ·IYYA HARAF·AAT SABAEIC ALPHABET

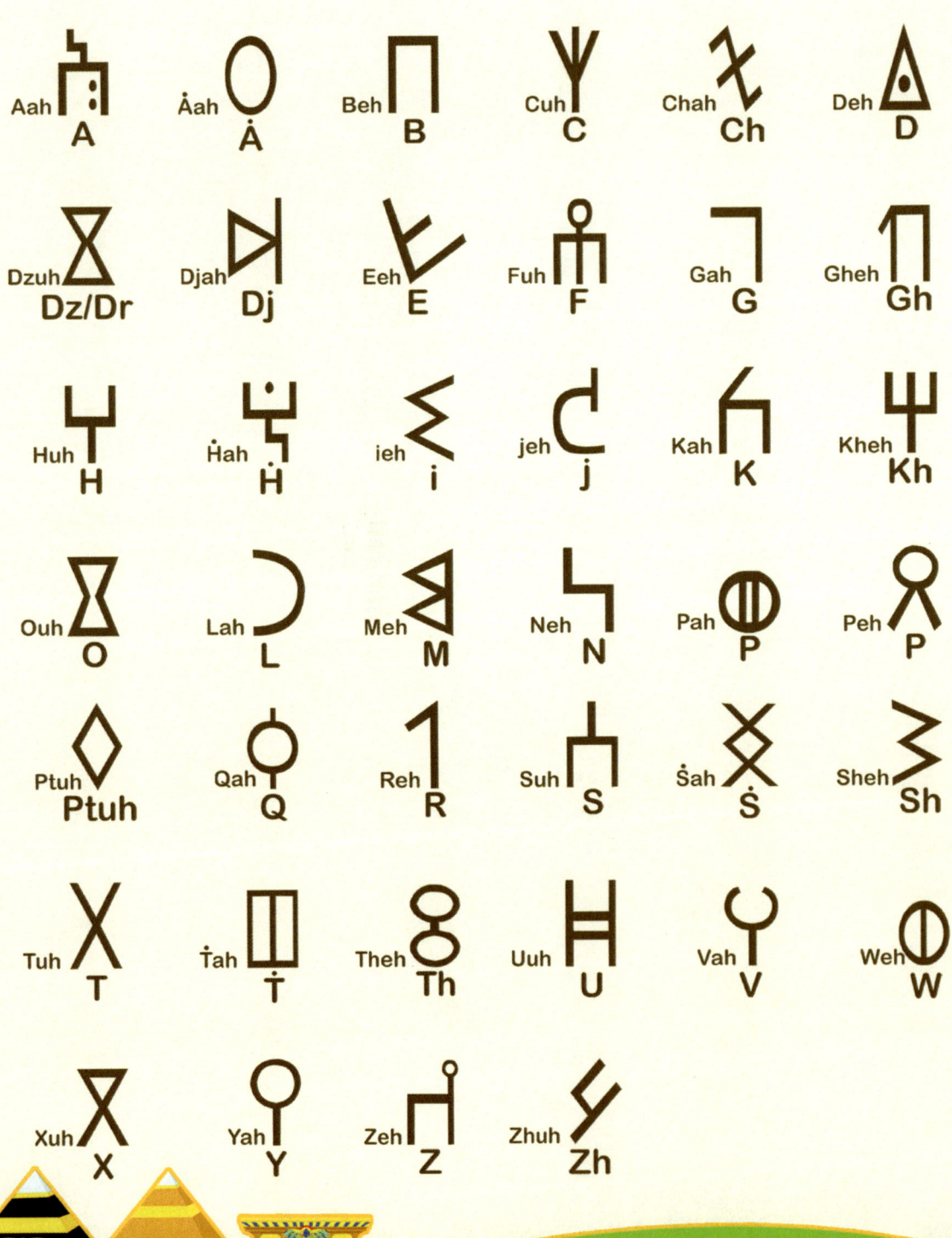

Aah **A**	Åah **Å**	Beh **B**	Cuh **C**	Chah **Ch**	Deh **D**
Dzuh **Dz/Dr**	Djah **Dj**	Eeh **E**	Fuh **F**	Gah **G**	Gheh **Gh**
Huh **H**	Ḣah **Ḣ**	ieh **i**	jeh **j**	Kah **K**	Kheh **Kh**
Ouh **O**	Lah **L**	Meh **M**	Neh **N**	Pah **P**	Peh **P**
Ptuh **Ptuh**	Qah **Q**	Reh **R**	Suh **S**	Ṡah **Ṡ**	Sheh **Sh**
Tuh **T**	Ṫah **Ṫ**	Theh **Th**	Uuh **U**	Vah **V**	Weh **W**
Xuh **X**	Yah **Y**	Zeh **Z**	Zhuh **Zh**		

ṠABAṪ·IYYA RAQAM·AAT SABAEIC NUMBERS

Symbol	Number	Name
⊙	0	Safu
I	1	Wahu
∧	2	Athu
⅄	3	Thalu
+	4	Rabu
✕	5	Khamu
✳	6	Satu
✴	7	Sabu
✳	8	Tamu
⊕	9	Tasu
∩	10	Ashu
∩I	11	Ashu Wu Wahu
∩∧	12	Ashu Wu Athu
∩⅄	13	Ashu Wu Thalu
∩+	14	Ashu Wu Rabu
∩✕	15	Ashu Wu Khamu
∩✳	16	Ashu Wu Satu
∩✴	17	Ashu Wu Sabu
∩✳	18	Ashu Wu Tamu
∩⊕	19	Ashu Wu Tasu
∩∩	20	Athura
∩∩I	21	Athura Wahu
∩∩∧	22	Athura Athu
∩∩⅄	23	Athura Thalu
∩∩+	24	Athura Rabu
∩∩✕	25	Athura Khamu
∩∩✳	26	Athura Satu
∩∩✴	27	Athura Sabu
∩∩✳	28	Athura Tamu
∩∩⊕	29	Athura Tasu
∩∩ (stacked)	30	Thalura
∩∩ (stacked)	40	Rabura
∩∩ (stacked)	50	Khamura
∩∩∩ (stacked)	60	Satura
∩∩∩ (stacked)	70	Sabura
∩∩∩ (stacked)	80	Tamura
∩∩∩ (stacked)	90	Tasura

Symbol	Name	Value
℮	Mayu	100
℮℮	Mayu Mayu	200
Afu	Afu	1,000
Ashu Afu	Ashu Afu	10,000
Mayu Afu	Mayu Afu	100,000
Malu	Malu	1,000,000

Lah'aj Kids MULTIPLICATION WORKSHEETS WITH ŚABATIYYA

/∩∩x · 25

RANAN · NAME

KHADUT · DATE

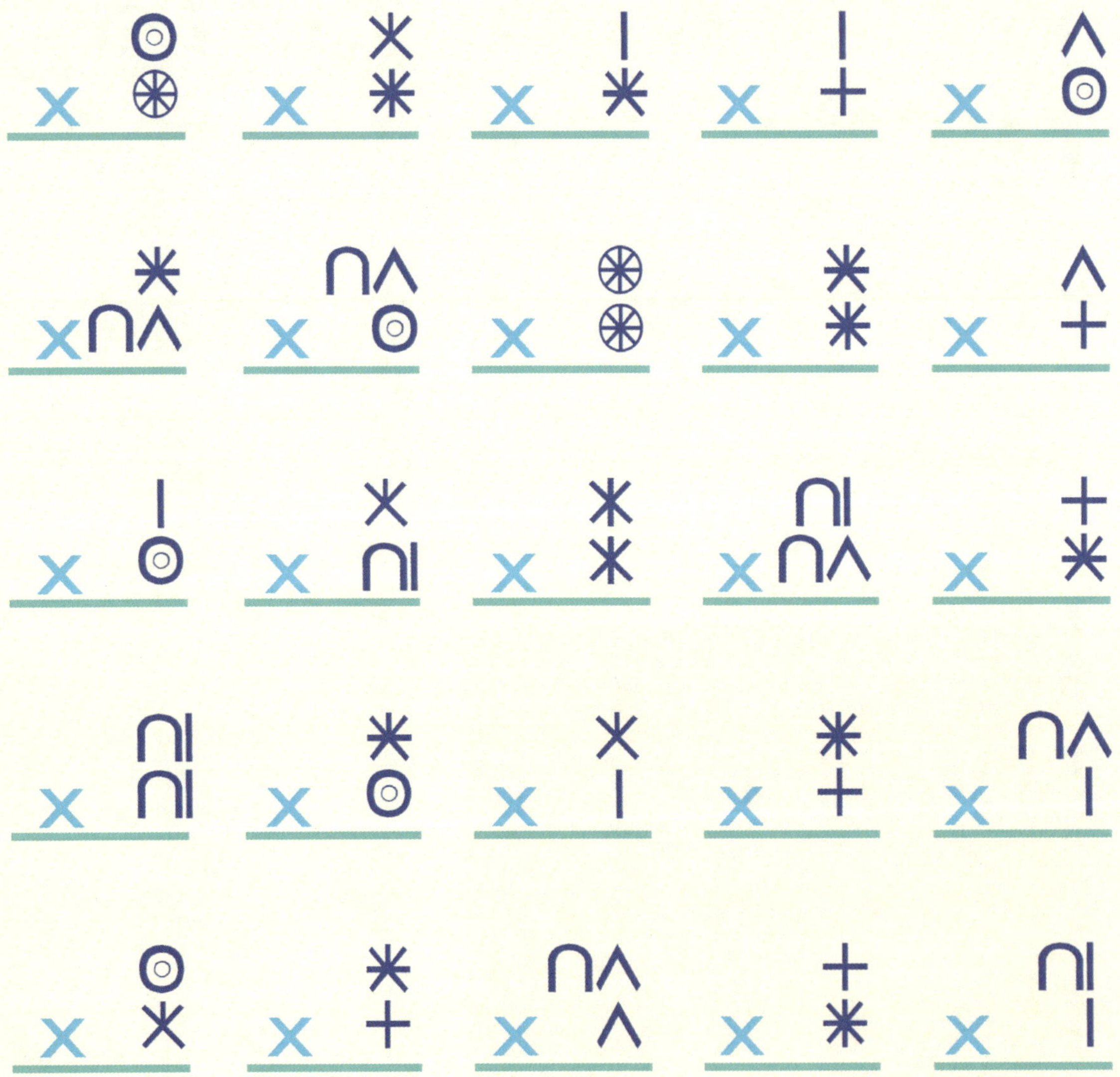

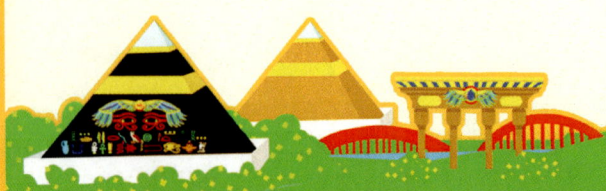

ᔭᐦᔭᐦꞋ·RANAN·NAME

᙭Ꮋᐊᔭᖻ·KHADUT·DATE

Lah'aj Kids

MULTIPLICATION WORKSHEETS WITH ŚABATIYYA

· 25

RANAN · NAME

KHADUT · DATE

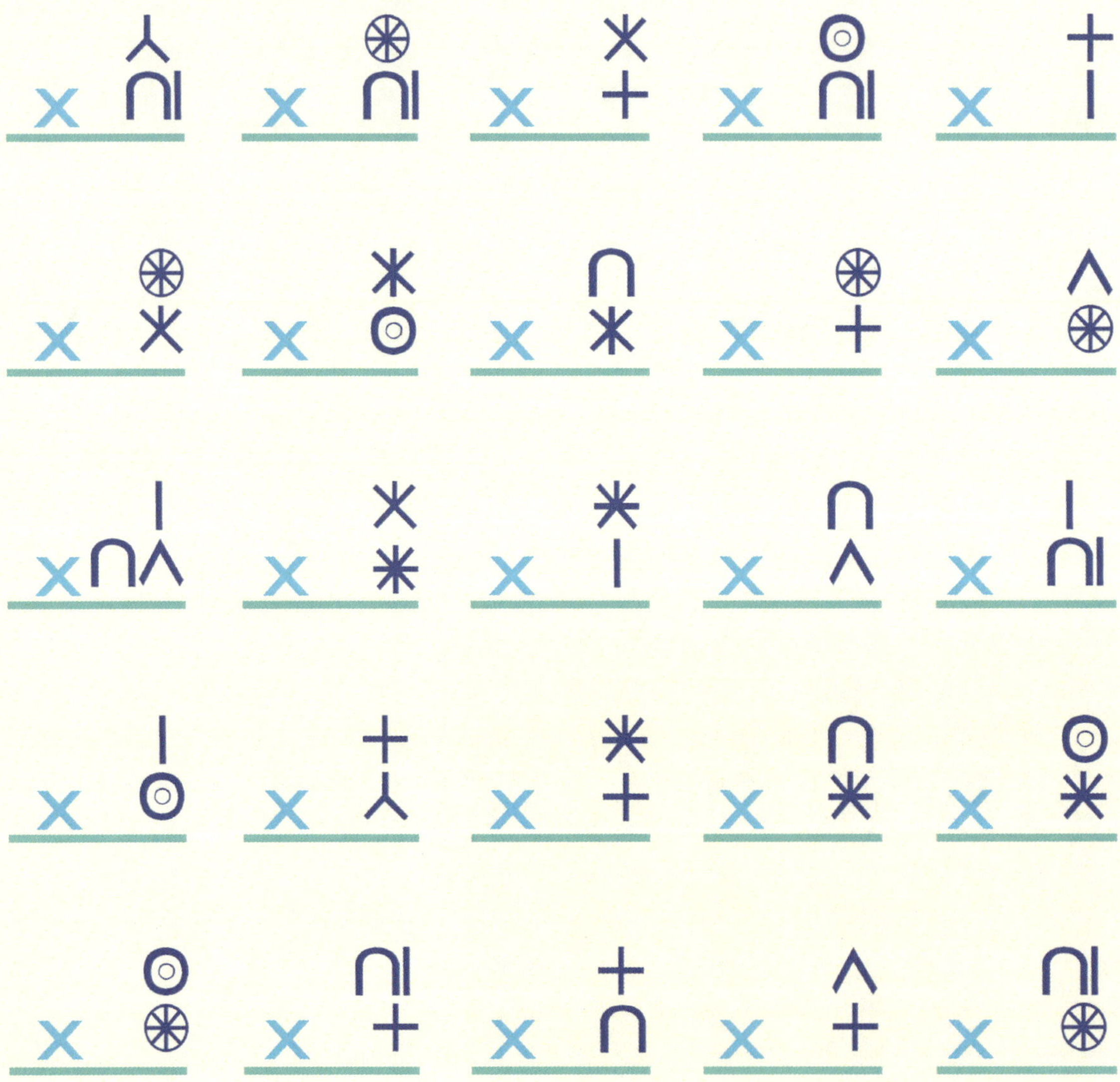

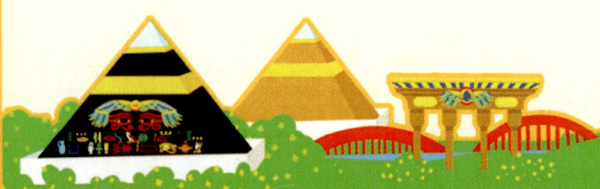

ᔕᐟᔕᐟ1·RANAN·NAME

᙭Ꮋᐃ♓ᙍ·KHADUT·DATE

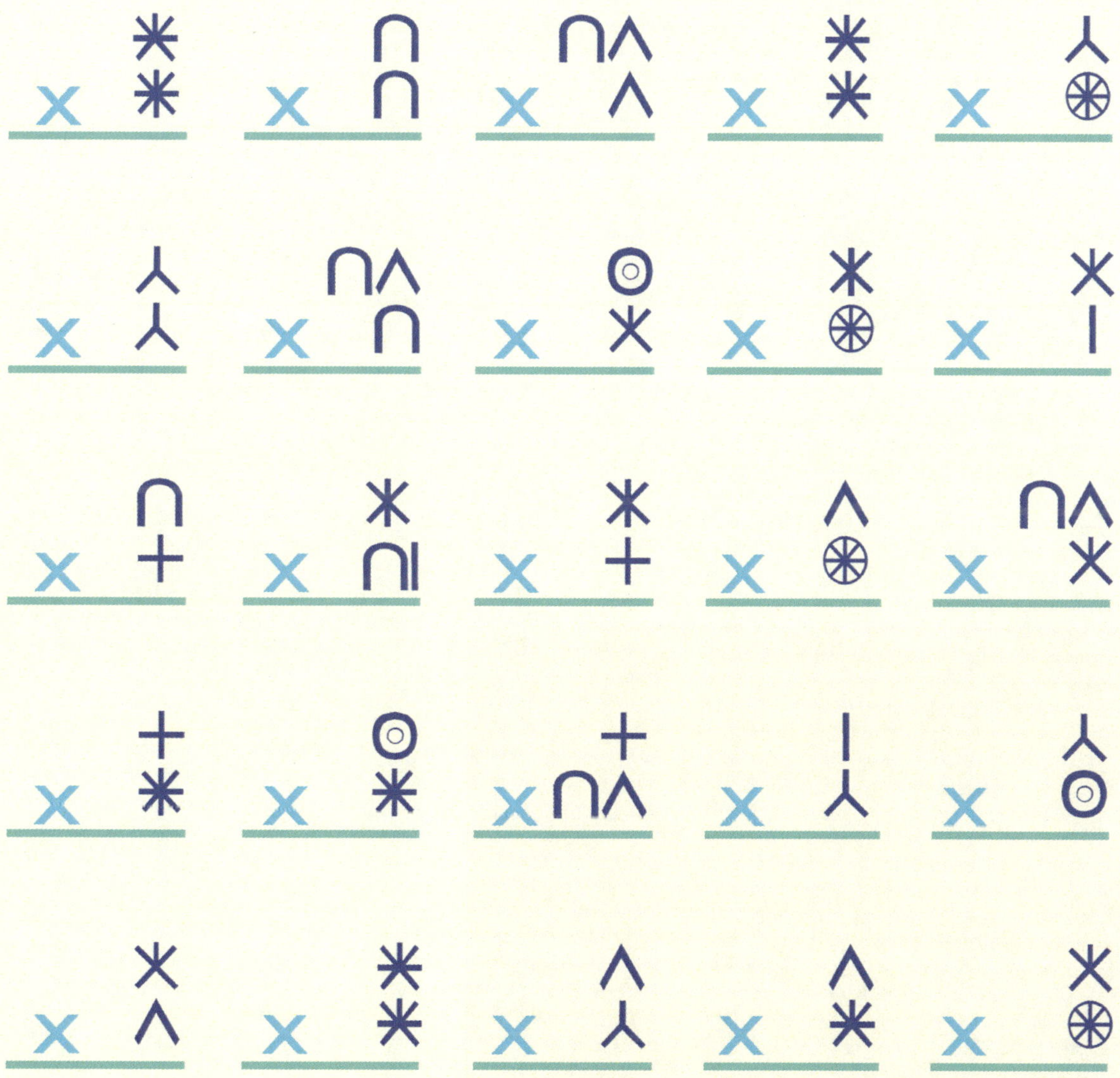

Lah'aj Kids

MULTIPLICATION WORKSHEETS WITH ŚABAṪIYYA

/∩∩x · 25

RANAN · NAME

KHADUT · DATE

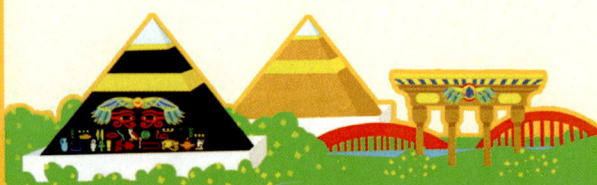

ᚷᚺᚷᚺᛏ·RANAN·NAME

ᚷᚺᚷᚺᛏ·KHADUT·DATE

∩∧ ✕ ✳	∩∐ ✕ ∩∧	✝ ✕ ∩∐	◎ ✕ ◎	∩∐ ✕ ∧⟩
✝ ✕ ✝	∩∧ ✕ ◎	✲ ✕ ✳	⅄ ✕ ✳	∩∐ ✕ ✳
∧ ✕ ✳	✲ ✕ ✲	┃ ✕ ✲	✳ ✕ ✳	⟩∧⟩ ✕ ⟩∧⟩
∩∐ ✕ ∩∐	◎ ✕ ∩	∧ ✕ ◎	∩✕ ✕ ∩∐	✝ ✕ ◎
┃ ✕ ┃	∩∐ ✕ ✳	✳ ✕ ✳✕	✲ ✕ ∩	✳✕ ✕ ✳✕

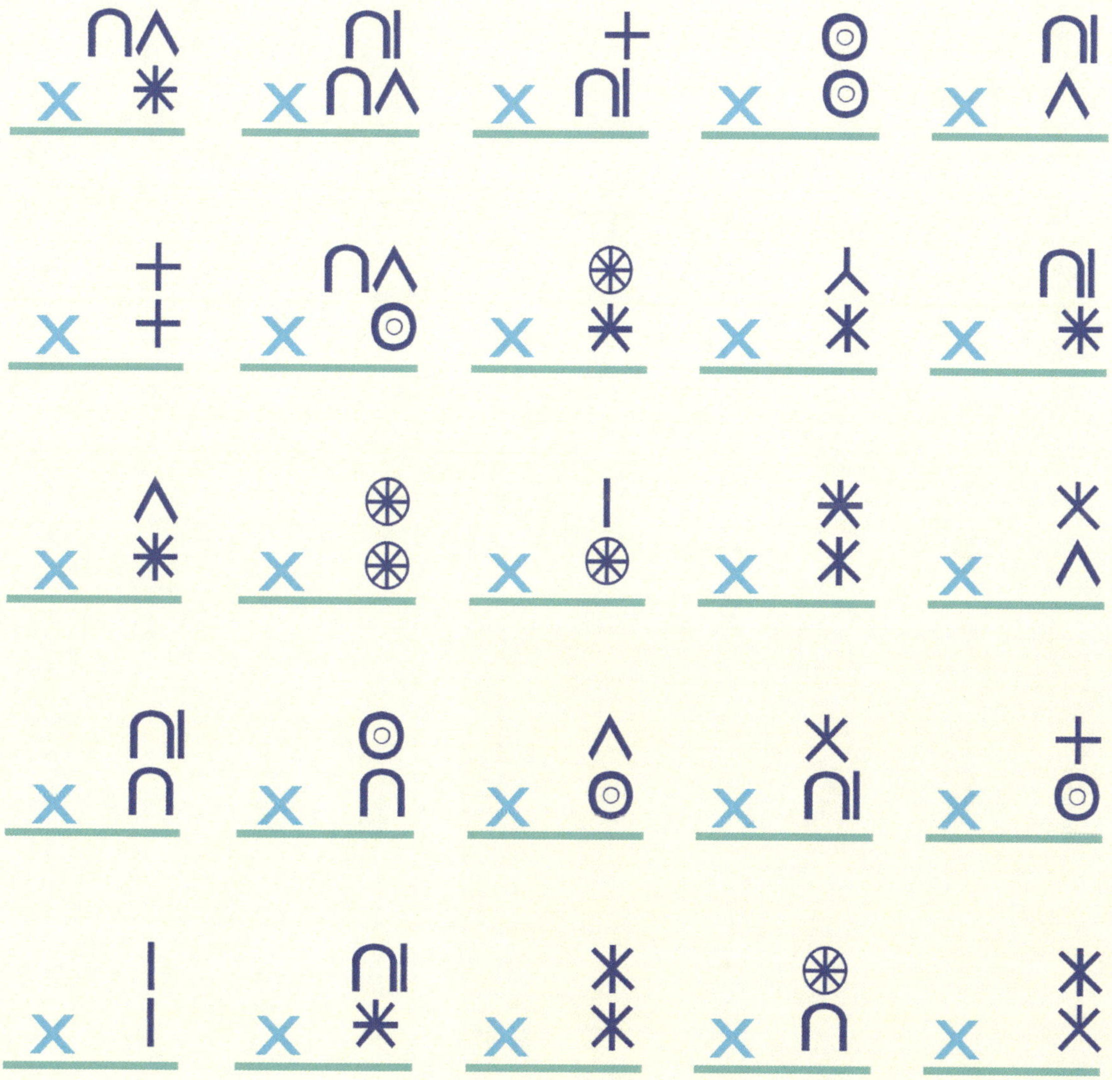

ꟾꙄꙄꞀꙄꞀ·RANAN·NAME

ΧꞀΔꙄꟼᛉ·KHADUT·DATE

× ∩| 　× ⊕ 　× ⊀ 　× ∩∧ 　× ⋏

× ⊙∩ 　× ∩⋇ 　× ∧⊙ 　× ∧⋇ 　× ⋇⋏

× ⋇∩ 　× ⋇⊙ 　× ⋏⋏ 　× ⊙∩ 　× |

× ⋏⋇ 　× ∧⊹ 　× ⊹⊕ 　× ∧⊕ 　× ∩∧∩

× |∩ 　× ⋏⋏ 　× ⋇∩ 　× ∧⋇ 　× ⊹⊹

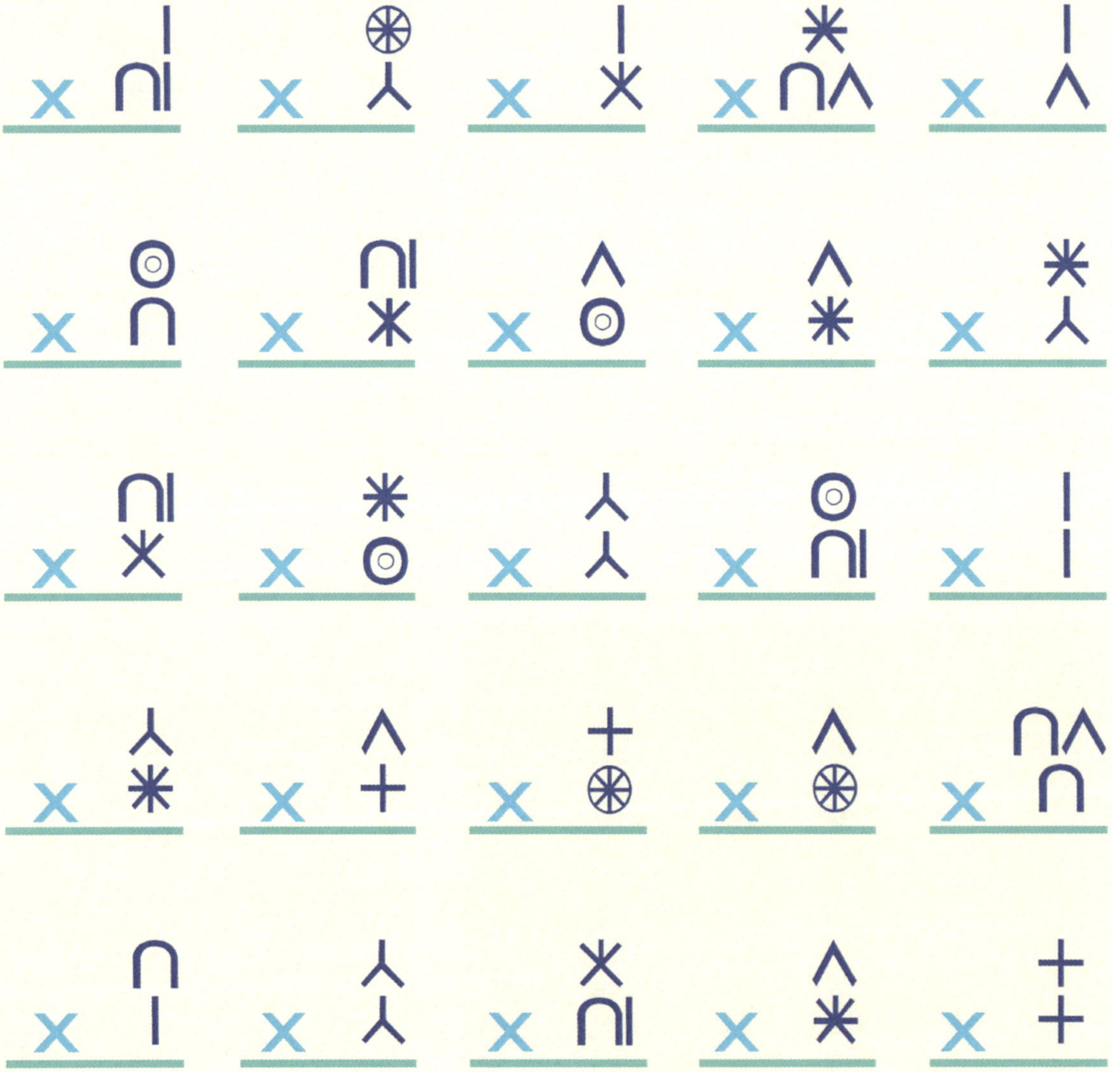

/∩∩⅂·25

ᔆᗷᔆᗷ1·RANAN·NAME

Хᖯᐱᔆᖯ·KHADUT·DATE

━ ━ ━ ━ ━ ━ ━ ━ ━ ━ ━
ᒡᐱᒡᐱ1·RANAN·NAME

━ ━ ━ ━ ━ ━ ━ ━ ━ ━ ━

᙭ᕼᐃᒣᛘ·KHADUT·DATE

ᒐᔐᒐᔐᒐ·RANAN·NAME

ⵝⴹⴸⵀⵞ·KHADUT·DATE

$$\times \begin{array}{c}\cap\wedge\\\cap\wedge\end{array} \qquad \times \begin{array}{c}\odot\\\divideontimes\end{array} \qquad \times \begin{array}{c}\wedge\\\divideontimes\end{array} \qquad \times \begin{array}{c}\divideontimes\\\divideontimes\end{array} \qquad \times \begin{array}{c}\circledast\\|\end{array}$$

$$\times \begin{array}{c}\curlyvee\\\dagger\end{array} \qquad \times \begin{array}{c}\odot\\|\end{array} \qquad \times \begin{array}{c}\divideontimes\\\divideontimes\end{array} \qquad \times \begin{array}{c}\cap\\\odot\end{array} \qquad \times \begin{array}{c}\wedge\\\odot\end{array}$$

$$\times \begin{array}{c}\cap\wedge\\\dagger\end{array} \qquad \times \begin{array}{c}\cap\sqcup\\\times\end{array} \qquad \times \begin{array}{c}\odot\\\divideontimes\end{array} \qquad \times \begin{array}{c}\cap\wedge\end{array} \qquad \times \begin{array}{c}\dagger\\\divideontimes\end{array}$$

$$\times \begin{array}{c}\divideontimes\\\cap\end{array} \qquad \times \begin{array}{c}|\\\dagger\end{array} \qquad \times \begin{array}{c}\divideontimes\\\divideontimes\end{array} \qquad \times \begin{array}{c}\wedge\\\circledast\end{array} \qquad \times \begin{array}{c}\divideontimes\\\curlyvee\end{array}$$

$$\times \begin{array}{c}\cap\\\circledast\end{array} \qquad \times \begin{array}{c}\divideontimes\\\divideontimes\end{array} \qquad \times \begin{array}{c}\circledast\\\circledast\end{array} \qquad \times \begin{array}{c}\dagger\\\divideontimes\end{array} \qquad \times \begin{array}{c}\wedge\\\cap\wedge\end{array}$$

Laĥaj Kids

MULTIPLICATION WORKSHEETS WITH ŚABAṬIYYA

/ПП⋇ · 25

▪▪▪▪▪▪▪▪▪▪

ᗺᕼᕼᕼ1 · RANAN · NAME

▪▪▪▪▪▪▪▪▪

ᕽᕼᗞᕼᗺ · KHADUT · DATE

$\times \sqcup\mathsf{I}$ $\times \curlyvee$ $\times \sqcap\divideontimes$ $\times \curlyvee$ $\times \sqcap\wedge$

$\times \wedge$ $\times \odot$ $\times \circledast$ $\times \sqcup\wedge$ $\times \sqcup\sqcup$

$\times \divideontimes$ $\times \odot$ $\times \divideontimes$ $\times \circledast$ $\times \curlyvee$

$\times \mathsf{I}$ $\times \sqcup\odot$ $\times \wedge$ $\times \curlyvee$ $\times \circledast$

$\times \uparrow$ $\times \sqcap\divideontimes$ $\times \sqcap\divideontimes$ $\times \sqcup\curlyvee$ $\times \divideontimes$

ᖴᕼᗝ⊕ ᗺᕽᕼ ᗺᕼᕽᕼᕽ · SAHAF ASHU WAHU · PAGE 11

ᒉᐊᒉᐊ1 · RANAN · NAME

Χᕼᐊᔭᖈ · KHADUT · DATE

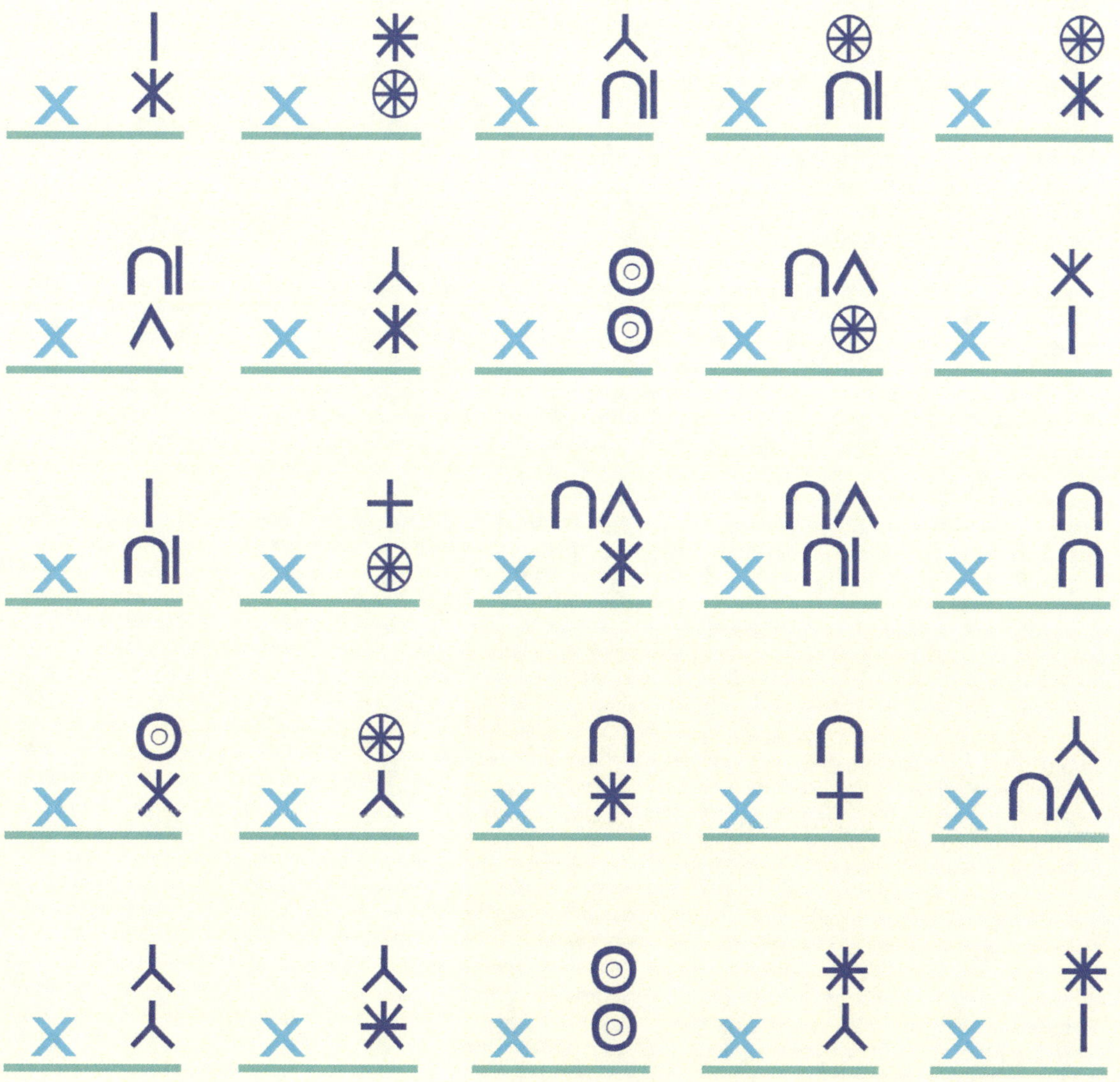

Laḩaj Kids

MULTIPLICATION WORKSHEETS WITH ṠABAṪIYYA

/ⴖⴖ⋇·25

ͽↄͽↄ1·RANAN·NAME

ⵝH△ꟼⵌ·KHADUT·DATE

/ΠΠ✕ · 25

ᚱᚪᚾᚪᚾ · RANAN · NAME

ᚷᚻᚪᚾᚤ · KHADUT · DATE

× 人

× ⅄ ⁕

× ⊙ ⁕

× ⁕ ⏐

× ⁕⁕

× ⊕ ⋂

× ⅄ ✕

× ⋂ ⁕

× ⊙ ✕

× ┼ ⅄

× ⊙ ⊕

× ⏐ ⋂

× ✕ ⁕

× ⊙ ⅄

× ⋂⅄ ⁕

× ┼

× ⁕ ⅄

× ⋂⅄ ⊙

× ⁕ ⊕

× ⊕ ⏐

× ⏐ ⋂

× ⋂ ⋂

× ⋂⅄ ⁕

× ⊙ ⊙

× ⊙ ⁕

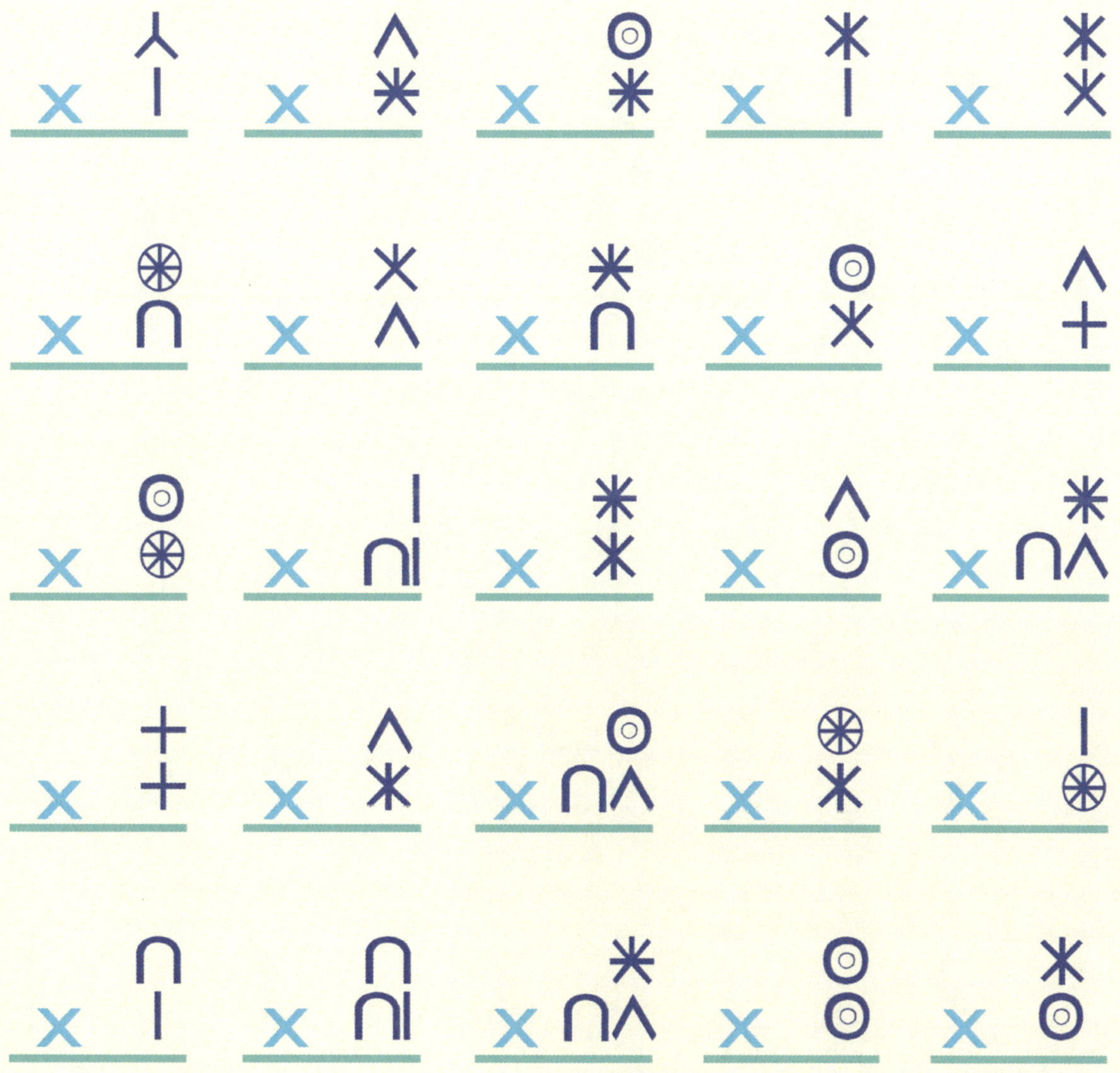

/∩∩✕·25

ᒐᕌᕌᒐ·RANAN·NAME

ᚷᚻᐃᚿᚿ·KHADUT·DATE

$$\times \begin{matrix} + \\ ✳ \end{matrix} \qquad \times \begin{matrix} ∪∧ \\ | \end{matrix} \qquad \times \begin{matrix} 丫 \\ ✳ \end{matrix} \qquad \times \begin{matrix} | \\ 丫 \end{matrix} \qquad \times \begin{matrix} ⊙ \\ + \end{matrix}$$

$$\times \begin{matrix} ∪ \\ ✳ \end{matrix} \qquad \times \begin{matrix} + \\ 丫 \end{matrix} \qquad \times \begin{matrix} ∪∧丫 \end{matrix} \qquad \times \begin{matrix} ∧ \\ ∪ \end{matrix} \qquad \times \begin{matrix} ∪∐ \\ ✳ \end{matrix}$$

$$\times \begin{matrix} ∪ \end{matrix} \qquad \times \begin{matrix} ∧ \\ ∩ \end{matrix} \qquad \times \begin{matrix} ✳ \\ | \end{matrix} \qquad \times \begin{matrix} ∧ \\ | \end{matrix} \qquad \times \begin{matrix} ∪∧ \\ 丫 \end{matrix}$$

$$\times \begin{matrix} ∪∐ \\ 丫 \end{matrix} \qquad \times \begin{matrix} 丫 \\ 丫 \end{matrix} \qquad \times \begin{matrix} ✳ \\ ✳ \end{matrix} \qquad \times \begin{matrix} | \\ ⊙ \end{matrix} \qquad \times \begin{matrix} 丫 \\ ✳ \end{matrix}$$

$$\times \begin{matrix} ∧ \\ ⊛ \end{matrix} \qquad \times \begin{matrix} ⊙ \\ 丫 \end{matrix} \qquad \times \begin{matrix} + \\ + \end{matrix} \qquad \times \begin{matrix} ✳ \\ ✳ \end{matrix} \qquad \times \begin{matrix} ∩∐ \\ | \end{matrix}$$

Laḥaj Kids
MULTIPLICATION WORKSHEETS WITH ṢABAṬIYYA

꠵ꠥꠥꠍ꠆·RANAN·NAME

ꠉꠖꠅꠍꠊ·KHADUT·DATE

× 人 ⊙   × ⊕⊕   × ∩ X   × ✳✳   × ⊕ ✚

× ✳ ✳   × ∩∧∩∪   × ✳ ✳   × ⊕ 人   × 人 ∧

× ✳ ✚   × ✳ ∩∪   × ∩∪   × X 人   × 人 ✳

× ⊕ ✳   × X ∩∪✳   × ✚ ✳   × ∩∧∩∧   × ∩∧ ✚

× ∩∪∧ ✚   × ✳ ∩∪∧   × ∧ ∧   × ✳ ✳   × ― ✳

ᔕᐁᔕᐁ1·RANAN·NAME

᙭ᐁᐃᐞᙎ·KHADUT·DATE

⋇ × ⋏	∏⋏ × ⋏	⋇ × ⊛	∏⋏ × ⋇	︱ × ⋇
⋇ × ∩	⋇ × ⋏	⋏ ×∩⋏	⋏ × ⋏	⊛ × ∩
∏⋏ × ∩	⋏ × ⋏	∩ × ⋏	⋏ × ︱	⋇ × ⊛
◉ × ◉	︱ × ∩	⋏ × ✝	⋇⋇ × ⋇⋇	⋇ ×∩⋏
∩ × ∩	⋇⋇ × ⋇⋇	⋇⋇ × ⋇⋇	∩ × ✝	⋇ × ∩

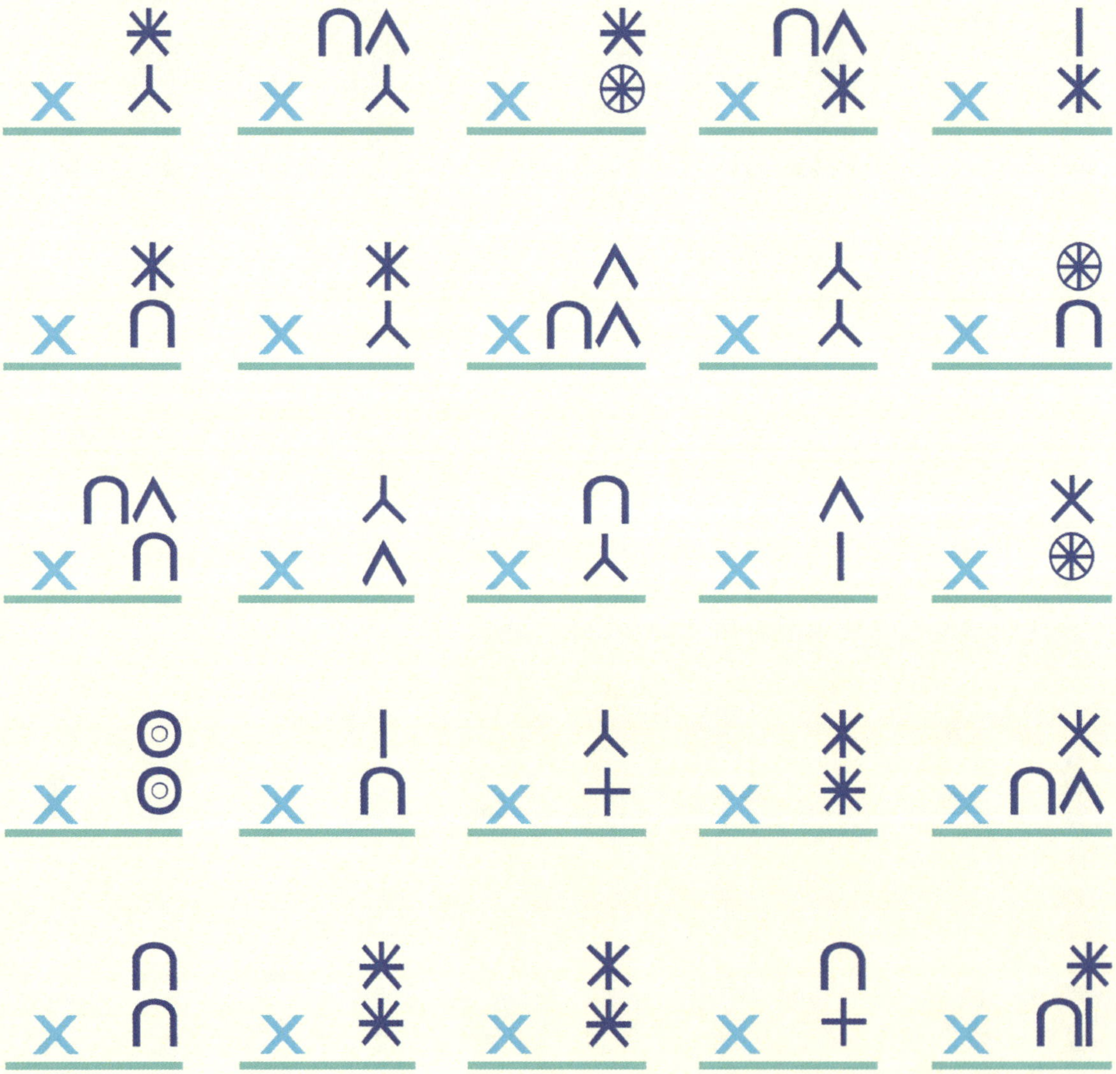

/ⴖⴖⵝ·25

᛭·RANAN·NAME

᛭·KHADUT·DATE

Lah'aj Kids
MULTIPLICATION WORKSHEETS WITH ŚABATIYYA

᚛ᚾᚾᚱᚾᚷᚔ·RANAN·NAME

ᚷᚺᚪᚾᚒᚔ·KHADUT·DATE

× ⊛	× ⋂	× ✳	× ⋂	× ✗
× ─	× ✳	× ✳	× ⋏	× ⋂⋀
× ✳	× ✳	× ◎	× ⋂	× ⋂
× ✳	× ⋏	× ⋀	× ✳	× ✛
× ✛	× ✳	× ✳	× ✳	× ◎

/ΠΠX · 25

ᚦᚨᚾᚨᚾ • RANAN • NAME

 Xᚼᚨᚾᚤ • KHADUT • DATE

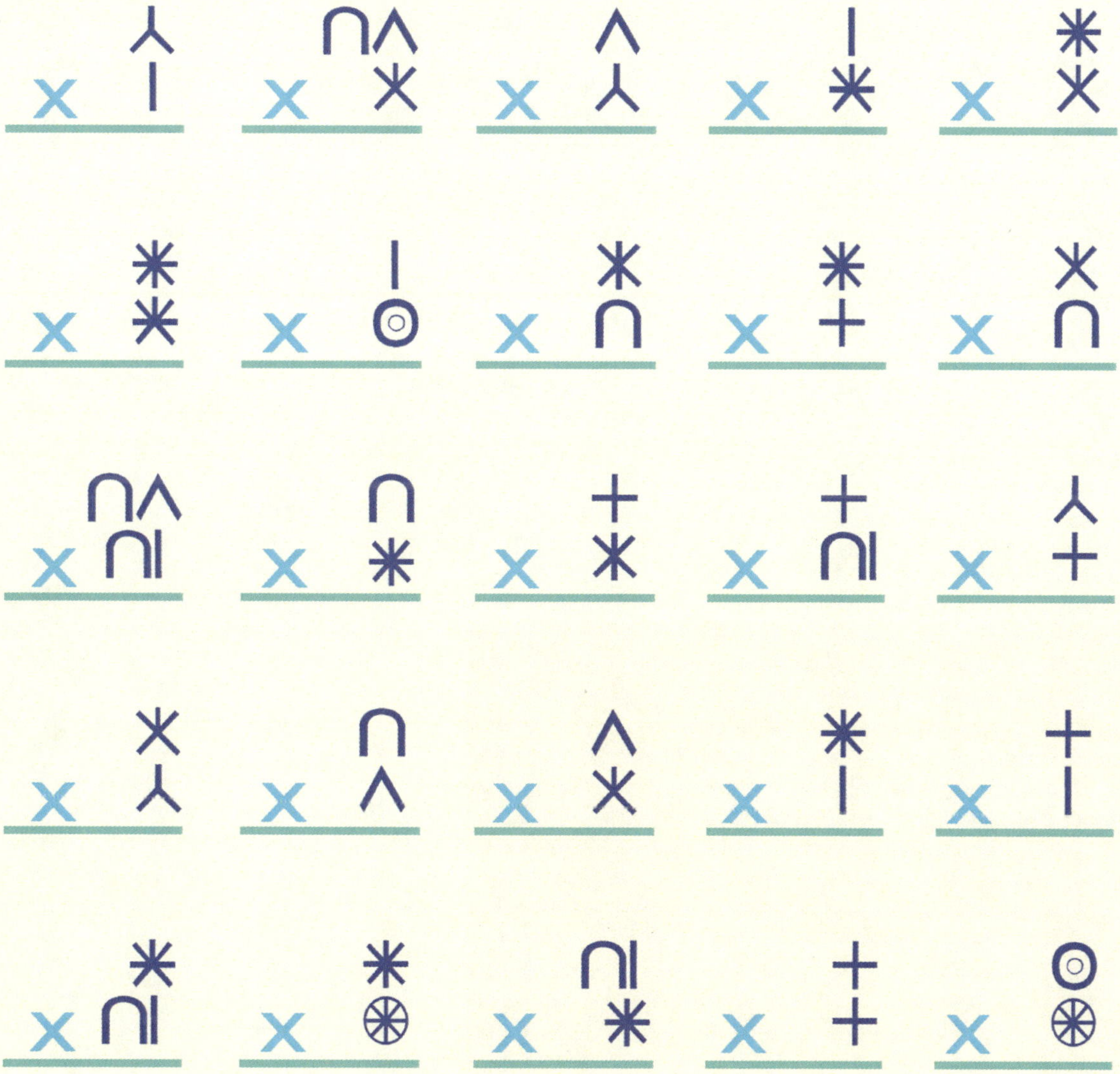

Laĥ'aj Kids

MULTIPLICATION WORKSHEETS WITH ŜABAṬIYYA

_____
ᔕᔕᔕ1·RANAN·NAME

_____
ᕼᕼᕼ·KHADUT·DATE

Laḱaj Kids

MULTIPLICATION WORKSHEETS WITH ŚABAṪIYYA

/ · 25

RANAN · NAME

KHADUT · DATE

ᚱᚨᚾᚨᚾ · RANAN · NAME

ᚲᚺᚨᛞᚢᛏ · KHADUT · DATE

Laĥ'aj Kids
MULTIPLICATION WORKSHEETS WITH ŚABAṬIYYA

ᚦᚼᚾᚼᚾ1·RANAN·NAME

ᚷᚼᚪᚾᚱ·KHADUT·DATE

NOTES

NOTES